ALL ABOUT BABY KOALAS

by Martha E. H. Rustad

PEBBLE
a capstone imprint

Pebble Emerge is published by Pebble, an imprint of Capstone.
1710 Roe Crest Drive
North Mankato, Minnesota 56003
www.capstonepub.com

Copyright © 2022 by Capstone. All rights reserved. No part of this publication may be reproduced in whole or in part, or stored in a retrieval system, or transmitted in any form or by any means, electronic, mechanical, photocopying, recording, or otherwise, without written permission of the publisher.

Library of Congress Cataloging-in-Publication Data
Names: Rustad, Martha E. H. (Martha Elizabeth Hillman), 1975- author.
Title: All about baby koalas / by Martha E.H. Rustad.
Description: North Mankato, Minnesota : Pebble, [2022] | Series: Oh baby! | Includes bibliographical references and index. | Audience: Ages 5-8 | Audience: Grades K-1 | Summary: "There's a new baby joining the colony. It's a koala joey! Learn all about baby koalas, including what they eat, what they weigh, how they're raised, and how big they grow"—Provided by publisher.
Identifiers: LCCN 2021002637 (print) | LCCN 2021002638 (ebook) | ISBN 9781663907912 (hardcover) | ISBN 9781663907882 (pdf) | ISBN 9781663907905 (kindle edition)
Subjects: LCSH: Koala—Infancy—Juvenile literature.
Classification: LCC QL737.M384 R87 2022 (print) | LCC QL737.M384 (ebook) | DDC 599.2/51392—dc23
LC record available at https://lccn.loc.gov/2021002637
LC ebook record available at https://lccn.loc.gov/2021002638

Image Credits
Shutterstock: Andras Deak, 17, apple2499, 9, artemiya, 21, Eric Isselee, back cover, Jen Watson, 16, Julia Nikitina, 20 top, Keitma, 11, Sacha M, 15, slowmotiongli, cover, 19, Susan Flashman, 10, thanongsuk harakunno, 13, Vichy Deal, 20 bottom, Zeyad Mohamed Edriss, 14, Superstock: D. Parer & E. Parer-Cook/Minden Pictures, 5, Suzi Eszterhas/Minden Pictures, 6

Editorial Credits
Editor: Alison Deering; Designer: Jennifer Bergstrom; Media Researcher: Tracy Cummins; Production Specialist: Tori Abraham

All internet sites appearing in back matter were available and accurate when this book was sent to press.

Table of Contents

A Tiny Joey 4

Leaving the Pouch 8

Koala Food 12

Life in the Trees 16

All Grown Up 18

    Make Your Mark 20

    Glossary 22

    Read More 23

    Internet Sites 23

    Index 24

Words in **bold** are in the glossary.

A TINY JOEY

Look at the baby koala! A baby koala is called a **joey**. It grew inside its mother for about one month. Then it crawled into a **pouch** on her belly.

A baby koala has no fur. It cannot hear or see. But it can hold on to its mother. She can open and close her pouch using a special muscle. She keeps her baby from falling out.

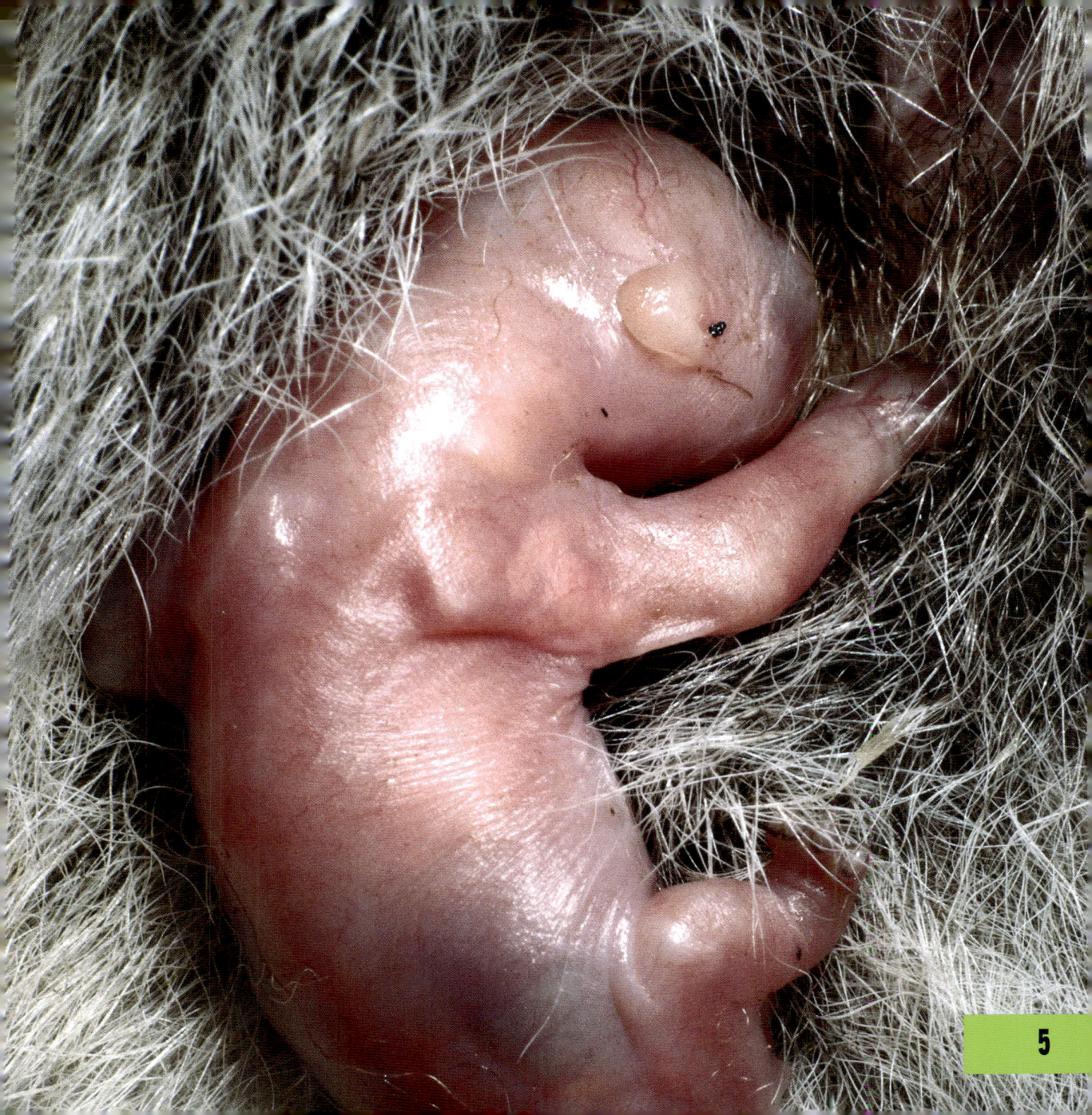

The baby koala lives in the pouch for about six months. It drinks milk from its mother's body.

The joey slowly gets bigger. Fuzzy fur grows on its body. Ears form on its head. Then it pokes its head out of the pouch. It opens its eyes and looks around. The pouch is still its home.

LEAVING THE POUCH

The baby keeps growing. Sometimes it gets out of the pouch. But it stays close to its mother. It has strong claws. They help the baby grip the fur on its mother's back.

The joey goes back to the pouch to drink milk. It cuddles close to its mother to sleep.

Joeys stay with their mothers in a small area. This is called a home **range**. Other koalas live nearby.

Koalas mark the trees in their areas. Their sharp claws scratch the bark. Males sometimes put their scent on the trees. This tells other koalas not to eat from them.

KOALA FOOD

Adult koalas eat the leaves of **eucalyptus** trees. But the leaves are hard for their stomachs to break down. Young koalas must get their bodies ready to eat this food.

Baby koalas eat **pap**. It comes from their mother's bottoms. Pap looks and smells like runny poop. But it will help the baby break down eucalyptus.

Koalas are picky eaters. Their black noses sniff leaves. They carefully pick which ones to eat. Mothers teach their young how to find the best leaves.

Koalas sleep about 20 hours each day. They wake up to eat at night. Adult koalas eat about 1 pound (0.5 kilograms) of leaves each night.

LIFE IN THE TREES

Koala joeys grow up in the trees. They do not even get down to drink. They get water from the plants they eat.

A growing joey starts to explore the trees. But sometimes it wanders too far away. Then it makes a yipping sound. Its mother comes and finds it.

As joeys get bigger, extra fur grows on their bottoms. This makes sleeping in trees more comfortable.

ALL GROWN UP

After about six months, a joey is too big for its mother's pouch. The joey rides on its mother's back for up to one year. After a year, it stops drinking milk. It eats only leaves.

Joeys leave their mother after one to two years. The joey finds its own range.

A female koala is fully grown by age 2.
A male koala is fully grown by 3 to 4.

MAKE YOUR MARK

Each koala has its own fingerprints, just like you do. Use your fingerprints to make a koala in its habitat.

What You Need

- paper
- markers or crayons
- gray paint
- a paper plate

What You Do

1. Draw a eucalyptus tree on the paper. Look back at the photos in this book to see what eucalyptus trees look like.

2. Pour a small amount of gray paint on the paper plate. Dip your fingers in the paint to make koalas in the trees. Your thumb can be its body, your pointer finger can be its head, and your pinkie can be its ears, legs, and arms.

3. Let the paint dry.

4. Add black eyes, a nose, and claws to your koala.

Glossary

eucalyptus (yoo-kuh-LIP-tuhs)—a kind of tree; the leaves are toxic to most animals

joey (JOH-ee)—a young koala

pap (PAP)—a substance eaten by baby koalas; it gets their bodies ready to eat eucalyptus

pouch (POWCH)—a flap of fur on a female koala; baby koalas grow inside the pouch

range (RAYNJ)—an area where an animal lives

Read More

Grodzicki, Jenna. *Baby Koalas*. Minneapolis: Bearport Publishing Company, 2021.

Kenah, Katharine. *Super Marsupials: Kangaroos, Koalas, Wombats, and More*. New York: HarperCollins Publishers, 2019.

Kras, Sara Louise. *Koalas: A 4D Book*. Mankato, MN: Capstone, 2019.

Internet Sites

San Diego Zoo Kids: Koala
kids.sandiegozoo.org/animals/koala

Koalas for Kids (and Grown Ups)
savethekoala.com/sites/savethekoala.com/files/uploads/koalakids.pdf

National Geographic Kids: 10 Ten Facts About Koalas!
natgeokids.com/uk/discover/animals/general-animals/ten-facts-about-koalas/

Index

adulthood, 19

appearance, 4, 7, 14

claws, 8, 10

diet, 7, 8, 12, 14, 15, 16, 18

eucalyptus, 12, 14–15, 18

fur, 4, 7, 8, 17

home ranges, 10, 18

newborns, 4

pap, 12

pouches, 4, 7, 8, 18

sleep habits, 8, 15, 17

sounds, 16

trees, 10, 12, 16, 17

water, 16